Guillermo Ricardo Valdez
María Susana Vecino
María Eugenia Pedrosa

The Didactic Triad in Trigonometry

Guillermo Ricardo Valdez
María Susana Vecino
María Eugenia Pedrosa

The Didactic Triad in Trigonometry

Concern and teaching occupation

ScienciaScripts

Cover image: www.ingimage.com

This book is a translation from the original published under ISBN 978-620-2-15926-5.

Publisher:
Sciencia Scripts
is a trademark of
Dodo Books Indian Ocean Ltd. and OmniScriptum S.R.L publishing group

120 High Road, East Finchley, London, N2 9ED, United Kingdom
Str. Armeneasca 28/1, office 1, Chisinau MD-2012, Republic of Moldova, Europe
Printed at: see last page
ISBN: 978-620-7-62517-8

Contents

PRESENTATION

In the framework of the Research Group: "Educational Research", as members of the Project called: *"Post-pandemic teaching practices in trained and in-training Mathematics teachers*" and as teachers in first year Mathematics courses at the Faculty, we will share the results obtained in the research carried out at the Faculty of Exact and Natural Sciences of the National University of Mar del Plata, regarding the knowledge and difficulties on basic aspects of Trigonometry of students taking Algebra in the first year. These students belong to the degree courses in Chemistry, Biochemistry, Physics and Mathematics. We have also collected data on the approach to the subject of Trigonometry taken by sixth-year secondary school teachers in their classes during the pandemic.

In addition, a didactic sequence related to the topic is presented for the sixth year of secondary school.

On the other hand, an experience is reported on a first year class of the Faculty, using GeoGebra, whose objective is to reaffirm concepts related to the properties of the modules and arguments of Complex Numbers (a subject in which concepts of Trigonometry are applied).

CHAPTER I

Valdez, Guillermo
Vecino, Susana

Our First Year College Algebra Students : trigonometry/diagnostics

Introduction

The present research aims to investigate how the shortcomings in the teaching of trigonometry in secondary school can cause disadvantages for students in the acquisition of knowledge and skills at the beginning of their respective university careers. In this case, it is about first year students of the Faculty of Exact Sciences of the UNMDP, entrants 2023, who studied their last years of secondary education in pandemic.

Authors such as Cantoral *et al.* (2015) point out that one of the fundamental aspects in the teaching of trigonometry lies in the simultaneous use of objects, concepts, processes and practices. Therefore, he assures that "the plurality of reference practices, their interaction with diverse contexts and the evolution of the individual's or group's own life, will re-signify the knowledge constructed up to that moment, enriching it with new meanings" (p. 15).

The secondary school teacher plays a fundamental role, as he/she is the one who facilitates the strategies that can guide the learning of trigonometry, which, being a knowledge prior to vector quantities, will benefit the continuity of the student's learning.

On the other hand, it is pertinent to reflect, for a quality mathematical production, on some of Dr. Claud^ Alsina's thoughts, which we share below:

Rational intelligence in fact presents two ways of thinking. Without going into the anatomical and physiological aspects of the left and right hemispheres of the human brain (Glennon, 1980) or into their specific specialisation or complementarity (Gazzaniga, 1985), we would like to recall here that on the one hand we have a way of thinking: verbal, gestural, logical, analytical, linear, sequential... with evident capacities for the identification of concepts, expression, step-by-step deduction, logical argumentation... But, at the same time, we have a way of thinking: visual-spatial, analogical, intuitive, synthetical, multiple and simultaneous processing, with capacities for seeing, communicating, relating, identifying structures, understanding metaphors, establishing analogies,

etc.
https://www.studocu.com/es-ar/document/universidad-de-buenos-aires/matematicas/alsina-claudi-claudi-apuntes-1-2/9878735 Retrieved 13-06-2023

In particular, in the context of the curricular designs of the 3rd year of Secondary Education in the Province of Buenos Aires, we begin with Trigonometric Ratios and the resolution of right triangles.

More precisely, the axis "Geometry and Magnitudes" includes Trigonometry.

The following objectives are then set out:

1. To know the trigonometric ratios of right triangles.
2. Use the scientific calculator to solve problems involving sides and angles of right triangles.
3. With the teacher's help, learn the cosine theorem and some applications.

Likewise, in the curricular designs of the 6th year of Secondary Education in the Province of Buenos Aires, in the Axis: Algebra and Functions, the synthetic nucleus: Trigonometric functions is exposed.

Then, make it explicit:

Trigonometric functions are used in science to describe periodic phenomena, which require their domains to be real numbers. For this reason, their study must be framed in the study of functions of B B.

The time devoted to the analysis and discussion of the scales chosen on the axes for graphing will allow us to review concepts of real numbers; as well as to distinguish this functional view from what has been studied in the resolution of triangles.

To solve trigonometric equations it is necessary to read from the unit circle, so as not to resort to reduction formulas that are not very clear to students.

Diagnostic Test

In particular, before starting Unit 2: **"Complex Numbers",** of the subject Algebra, which is studied by first year students of our Faculty, a diagnostic test was carried out to corroborate the level of knowledge about **"Trigonometry"** in our students belonging to the degree courses in Chemistry and Biochemistry, and Teaching and Bachelor's Degree in Physics and Teaching and Bachelor's Degree in Mathematics. This subject is fundamental and prerequisite to be able to express and operate with complex numbers.

The test consists of 5 items:

Item 1 consists of five questions with statements in which you have to determine their truth or falsity with the corresponding justification of the chosen answer. This item aims to assess students' knowledge of the possible values that the functions sine, cosine and tangent can take.

Items 2 and 3 are intended to assess the transition from circular to sexagesimal systems and vice versa.

The objective of item 4 is to calculate the sin(") where κ is the angle determined by the positive real axis of "x" and the terminal side passing through a given point.

In item 5, students are asked to comment on their knowledge of Trigonometry (open question).

The test statements were as follows:

Diagnostic Activities

The following activities aim to review trigonometric concepts that are necessary to deal with complex numbers in polar form.

1) If cr is an angle. Rate as true or false (T or F) the following statements and justify them:

	Enunclados	Vo F	Justification
	Enunciados	**V o F**	**Justificación**
a)	Para algún α, $sen\ \alpha = \frac{3}{2}$		
b)	Para algún α, $tang\ \alpha = 15$		
c)	Para algún α, $-\cos \alpha = 2$		
d)	Para ningún α, $sen\ \alpha = \sqrt{2}$		
e)	Para ningún α, $sen\ \alpha = \frac{\sqrt{5}}{4}$		

2) To express the angle in the circular system, $\beta = 75° + 360°.k \quad con\ k = 7$

3) Express in the sexagesimal system $\frac{26}{5}\pi$ y give its congruent in the first rotation.

4) The terminal side of an angle y referred to a Cartesian coordinate system contains the point P(-2,3)- Calculate $sin\ \gamma$ (give the result with rationalised denominator).

5) Te would like to ask you to comment on this issue.

The resolution of the diagnostic test

Attached is the resolution of exercise 1 by one of the students and the resolution of Rems 2, 3 and 4 by another student, as none of the participants completed the diagnostic test completely without errors.

1) Let *a* be an angle. Classify the following statements as true or false (T or F) and justify

1) Sea α un ángulo. Calificar de verdaderas o falsas (V o F) las siguientes afirmaciones y justificar

	Enunciados	V o F	Justificación
a)	Para algún α, $sen\,\alpha = \frac{3}{2}$	F.	La imagen de la función seno es [-1,1]. y 3/2 > 1.
b)	Para algún α, $tang\,\alpha = 15$	V.	La imagen de la función tangente involucra a todos los números reales.
c)	Para algún α, $-\cos\alpha = 2$	F.	cos α = -2 es imposible, ya que la imagen de la función coseno es [-1,1].
d)	Para ningún α, $sen\,\alpha = \sqrt{2}$	V.	|√2| > 1, y la imagen de la función seno es [-1,1].
e)	Para ningún α, $sen = \frac{\sqrt{5}}{4}$	F.	|√5/4| < 1 (0,34...) y la imagen de la función seno es [-1,1].

2) Expresar en el sistema circular el ángulo $\beta = 75° + 360°.k$ con $k = 7$.

$\beta = 75° + 360°.7 = 2595°$

360 — 2π
2595 → x = $\frac{173}{12}\pi$ ✓

3) Expresar en el sistema sexagesimal el ángulo $\frac{26}{5}\pi$ y dar su congruente en el primer giro.

360° → 2π
936° — $\frac{26}{5}\pi$

936° ∟360°
−720 2
216

936° ≡ 216° ✓

4) El lado terminal de un ángulo y referido a un sistema de coordenadas cartesianas contiene al punto P(-2,3).
Calcular sen y (dar el resultado con denominador racionalizado).

$z^2 = 2^2 + 3^2$
$z^2 = 13$
$|z| = \sqrt{13} \to z = \sqrt{13}$

$sen\,y = \frac{3}{\sqrt{13}} \cdot \frac{\sqrt{13}}{\sqrt{13}} = \frac{3\sqrt{13}}{13}$

$sen\,y = \frac{3\sqrt{13}}{3}$ ✓

5) Te pedimos que ahora nos expreses algún comentario respecto a tus conocimientos previos sobre Trigonometría

Mis conocimientos sobre trigonometría son "buenos" pero me cuesta relacionar los datos.

Some Results:

The diagnostic test was taken by 38 students from the Biochemistry and Chemistry degree courses. The following table shows the **results obtained:**

Exercise 1	Answer Good	Answer Wrong	Unresolved	Justification		
				Well	Mal	Unresolved
a)	24	1	13	18	2	18
b)	20	2	16	13	1	24
c)	21	0	17	15	1	22
d)	16	5	17	14	1	23
e)	14	6	18	15	1	22

Exercises	**2**			**3**				**4**			
	Well	Mal	Without Resolve	Well	Reg.	Mal	Without Resolve	Well	Reg.	Mal	Without Resolve

	13	8	17	13	8	2	15	7	1	4	26

Percentage

Exercise 1	Answer Good	Answer Wrong	Without Resolve	Justification		
				Well	Mal	Without Resolve
a)	63,16%	2,63%	34,21%	47,37%	5,26%	47,37%
b)	52,63%	5,26%	42,11%	34,21%	2,63%	63,16%
c)	55,26%	0%	44,74%	39,48%	2,63%	57,89%
d)	42,10%	13,16%	44,74%	36,84%	2.63%	60,53%
e)	36,84%	15,79%	47,37%	39,48%	2,63%	57,89%

Exercise	2		
	Well	Mal	Unresolved
	34,21%	21,05%	44,74%

Exercise	3			
	Well	Reg.	Mal	Unresolved
	34,21%	21,05%	5,26%	39,48%

Exercise	4			
	Well	Reg.	Mal	Unresolved
	18,42%	2,63%	10,53%	68,42%

The diagnostic test was taken by 34 students from the following degree courses: Profesorado y Licenciatura en F^sica and Profesorado y Licenciatura en Matematica. The following table shows the **results obtained**:

Exercise 1	Answer Good	Answer Wrong	Unresolved	Justification		
				Well	**Mal**	**Unresolved**
a)	22	8	4	9	8	17
b)	22	8	4	8	4	22
c)	21	5	8	6	7	21
d)	13	12	9	7	6	21
e)	10	13	11	8	3	23

Exercises	2			3				4			
	Well	Mal	Without Resolve	Well	Reg.	Mal	Without Resolve	Well	Reg.	Mal	Without Resolve
	4	13	17	10	3	7	14	5	2	10	17

Percentage

Exercise 1	Answer Good	Answer Wrong	Unresolved	Justification		
				Well	Mal	Unresolved
a)	64,71%	23,53%	11,76%	26,47%	23,53%	50,00%
b)	64,71%	23,53%	11,76%	23,53%	11,76%	64,71%
c)	61,76%	14,71%	23,53%	17,65%	20,59%	61,76%
d)	38,24%	35,29%	26,47%	20,59%	17,65%	61,76%
e)	29,41%	38,24%	32,35%	23,53%	8,82%	67,65%

Exercise	2		
	Well	Mal	Unresolved
	11,76%	38,24%	50%

Exercise	3			
	Well	Regular	Mal	Unresolved
	29,41%	8,82%	20,59%	41,18%

Exercise	4			
	Well	Regular	Mal	Unresolved
	14,71%	5,88%	29,41%	50%

As can be seen in the same diagnostic test in item 5, we asked the students to comment on the topic:

To analyse these responses, we considered all groups in a unified form: that is, 72 people.

5) We now ask you to comment on your previous knowledge of Trigonometry:

Answers item 5	
I know the basics of trigonometry	20
I saw the issue at the Secondary School	12
I saw the topic at the Faculty Entrance	9
I never saw the issue	7
I am not clear on the concepts	4
I learned the topic from Youtube	3
I had no maths lessons in 6th grade.	2
I remember little	2
No reply	13

Word cloud representing the answers obtained in rtem 5

Error analysis:

In exercise 1, looking at the tables of results, it can be seen that a large percentage of the students answer wrongly or do not know the variation of the trigonometric ratios sine, cosine and tangent.

Some of the productions in particular stand out:

1.a) There are cases of students who admit that for some α, $sen(\alpha)=\frac{3}{2}$, which is false. Moreover, in the justification they present, the geometrical interpretation is totally wrong, given that they interpret that in a right triangle a leg can measure 3 units and the hypotenuse 2, which shows their lack of knowledge of equivalent fractions.

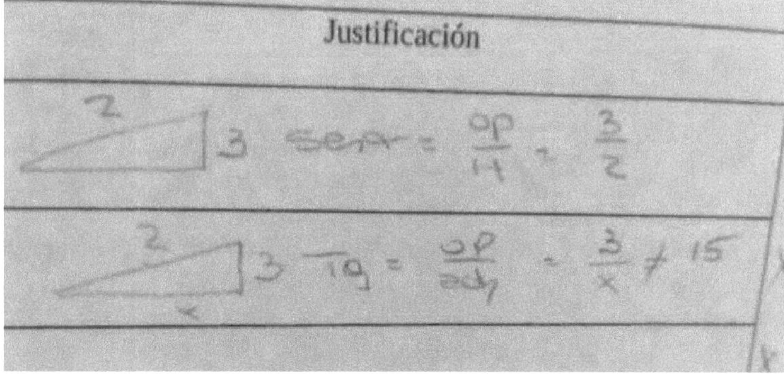

In exercises 2) and 3) Ignorance of the conversion from the sexagesimal system to the circular system and vice versa.

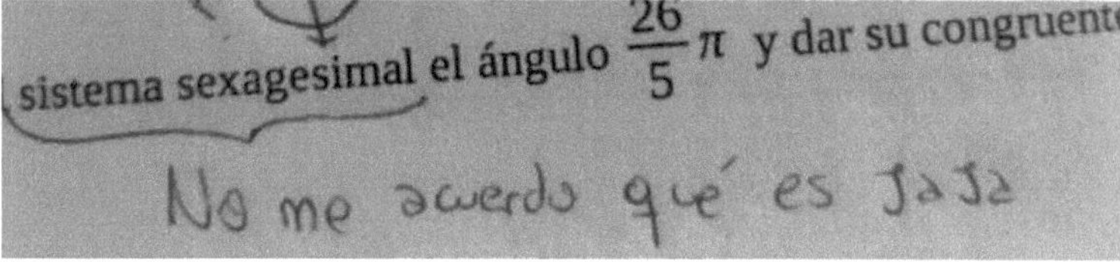

Also in exercise 3), there are cases of students who, although they calculated the amplitude of the angle correctly, could not find the congruent angle smaller than one turn.

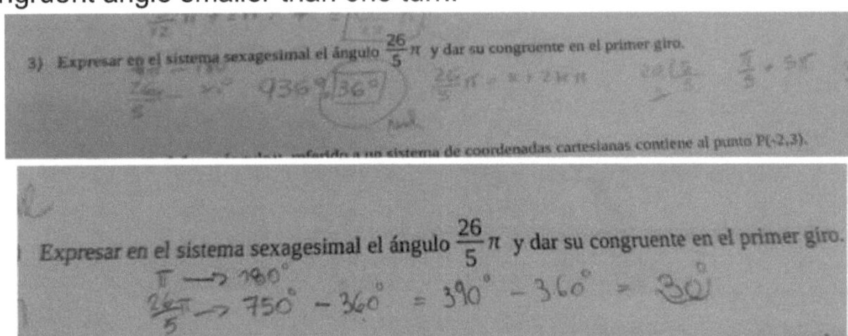

In exercise 4) some students had difficulty in representing the problem graphically, so that they were unable to locate the point P= (-2,3) in the Cartesian plane:

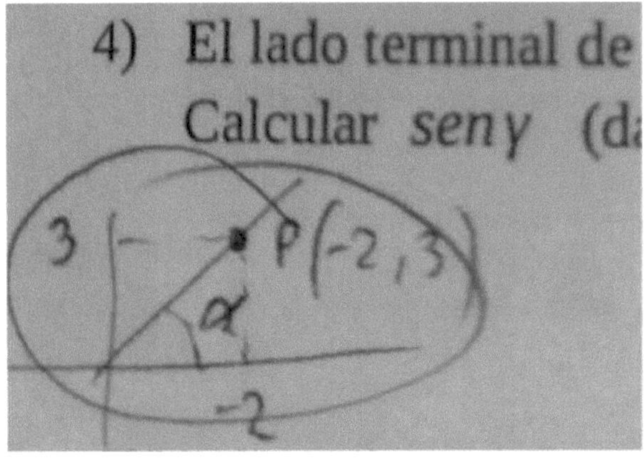

Other students, while correctly graphing the point P=(-2,3), misinterpret sin(γ) and confuse it with tg( ). γ

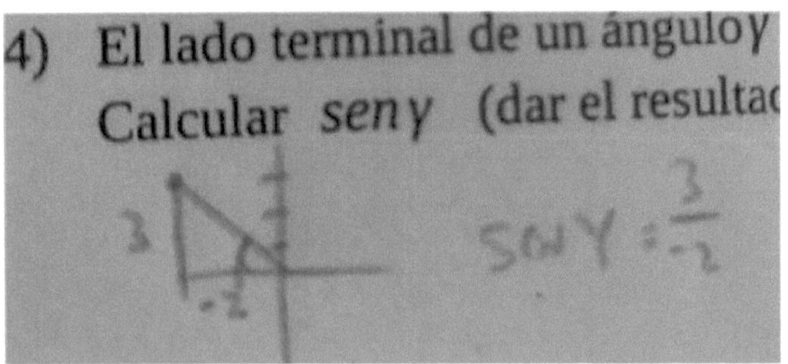

With regard to exercise 5) some students state that they have never seen anything about Trigonometry and others believe they have a good knowledge of Trigonometry, but indicate that they find it difficult to relate the data in the problems presented to the subject.

4) El lado terminal de un ángulo γ referido a un sistema de coordenadas cartesianas contiene al punto P(-2,3). Calcular sen γ (dar el resultado con denominador racionalizado).

Te pedimos que ahora nos expreses algún comentario respecto a tus conocimientos previos sobre Trigonometría

mis conocimientos sobre trigonometría son "buenos" pero me cuesta relacionar los datos

...dimos que ahora nos expreses algún comentario respecto a tus conocimientos previos sobre ...ometría

NUNCA VI NADA DE TRIGONOMETRÍA HASTA AHORA

The results show that there is little or no knowledge of "Trigonometry", which causes serious difficulties for the development of content corresponding to subjects such as linear algebra, calculus of one variable, calculus of several variables, physics, etc. Every teacher of Mathematics in Secondary School recognises that the treatment of this subject requires:

Understand the concepts, practice them, apply them, transfer them to other areas of knowledge and even to everyday life. To achieve this, it is necessary to have adequate time, which clearly did not exist during the pandemic.

For this reason, it was necessary to incorporate in the classroom of the Moodle platform that we use in the subject as a repository of the material, videos on basic concepts of trigonometry as well as complementary material and to add an "inverted class" on the subject to facilitate the understanding of the different representations of a complex number, a subject that has as a prerequisite concepts of Trigonometry.

In this sense, a table with the values of the trigonometric functions of the notable angles in all quadrants of the trigonometric circle was added to the material (see annex), material and videos that worked in IAM (Introduction to Mathematics) on the use of the calculator / telephone to obtain the sine and cosine of angles and a GeoGebra applet programmed by a teacher of the Department of Mathematics to reaffirm theoretical concepts concerning the properties of module and argument of complex numbers.

Finally, and as additional data, we include the tables referring to the totals and percentages of students who passed, failed, dropped out, had the opportunity to make up mid-term exams and those who did not take any of the mid-term exams (free) for the different groups of students during the first four-month period of 2023.

Career	No.of registered	Approved	Unsuccessful	Abandonments	Free
Bioqwmica	121	44	36	22	19
Chemistry	30	6	5	14	5
Mathematics	27	7	4	10	6
F^sica	42	15	10	11	6

Career	Number of registered participants	Approved% Approved	Unapproved% Unapproved% Unapproved% Unapproved% Unapproved% Unapproved% Unapproved	Abandonments% Abandonments% Abandonments% Abandonments	Free% Free
Bioqwmica	121	36,36	29,75	18,18	15,70
Chemistry	30	20,00	16,67	46,67	16,67
Mathematics	27	25,93	14,81	37,04	22,22
F^sica	42	35,71	23,81	26,19	14,29

ANNEX: TABLE OF TRIGONOMETRIC FUNCTIONS OF THE NOTABLE ANGLES

	0°	30°	45°	60°	90°	120°	135°	150°	180°	210°	225°	240°	270°	300°	315°	330°
θ	0	$\frac{\pi}{6}$	$\frac{\pi}{4}$	$\frac{\pi}{3}$	$\frac{\pi}{2}$	$\frac{2\pi}{3}$	$\frac{3\pi}{4}$	$\frac{5\pi}{6}$	π	$\frac{7\pi}{6}$	$\frac{5\pi}{4}$	$\frac{4\pi}{3}$	$\frac{3\pi}{2}$	$\frac{5\pi}{3}$	$\frac{7\pi}{4}$	$\frac{11\pi}{6}$
sen θ	0	$\frac{1}{2}$	$\frac{\sqrt{2}}{2}$	$\frac{\sqrt{3}}{2}$	1	$\frac{\sqrt{3}}{2}$	$\frac{\sqrt{2}}{2}$	$\frac{1}{2}$	0	$-\frac{1}{2}$	$-\frac{\sqrt{2}}{2}$	$-\frac{\sqrt{3}}{2}$	-1	$-\frac{\sqrt{3}}{2}$	$-\frac{\sqrt{2}}{2}$	$-\frac{1}{2}$
cos θ	1	$\frac{\sqrt{3}}{2}$	$\frac{\sqrt{2}}{2}$	$\frac{1}{2}$	0	$-\frac{1}{2}$	$-\frac{\sqrt{2}}{2}$	$-\frac{\sqrt{3}}{2}$	-1	$-\frac{\sqrt{3}}{2}$	$-\frac{\sqrt{2}}{2}$	$-\frac{1}{2}$	0	$\frac{1}{2}$	$\frac{\sqrt{2}}{2}$	$\frac{\sqrt{3}}{2}$
tan θ	0	$\frac{\sqrt{3}}{3}$	1	$\sqrt{3}$	ind	$-\sqrt{3}$	-1	$-\frac{\sqrt{3}}{3}$	0	$\frac{\sqrt{3}}{3}$	1	$\sqrt{3}$	ind	$-\sqrt{3}$	-1	$-\frac{\sqrt{3}}{3}$
csc θ	ind	2	$\sqrt{2}$	$\frac{2\sqrt{3}}{3}$	1	$\frac{2\sqrt{3}}{3}$	$\sqrt{2}$	2	ind	-2	$-\sqrt{2}$	$-\frac{2\sqrt{3}}{3}$	-1	$-\frac{2\sqrt{3}}{3}$	$-\sqrt{2}$	-2
sec θ	1	$\frac{2\sqrt{3}}{3}$	$\sqrt{2}$	2	ind	-2	$-\sqrt{2}$	$-\frac{2\sqrt{3}}{3}$	-1	$-\frac{2\sqrt{3}}{3}$	$-\sqrt{2}$	-2	ind	2	$\sqrt{2}$	$\frac{2\sqrt{3}}{3}$
cot θ	ind	$\sqrt{3}$	1	$\frac{\sqrt{3}}{3}$	0	$-\frac{\sqrt{3}}{3}$	-1	$-\sqrt{3}$	ind	$\sqrt{3}$	1	$\frac{\sqrt{3}}{3}$	0	$-\frac{\sqrt{3}}{3}$	-1	$-\sqrt{3}$

CHAPTER II

Pedrosa, Maria Eugenia
Palauro, Luda
Vecino, Susana

Teaching trigonometry in upper secondary school and its influence on the first year of university: teachers' perspectives and the effects of the pandemic.

According to Miguel de Guzman et. al (1988): ...it is necessary to treat in greater detail those subjects that clearly, in the general opinion of teachers, deserve special emphasis. Among these we can point out: statistics, probability, geometry and trigonometry. In our case, with emphasis on trigonometry, we recall that this part of mathematics was born around the 2nd century B.C., with Hipparchus' attempt to make astronomical observations a more exact art. Navigation, surveying, cartography... were other sources of motivation for the development of plane and spherical trigonometry. For the modern world, the following observation was even more important and triggered a new and fruitful development: the sine and cosine functions are periodical, i.e. as for all x they are verified:

$$sen(x + 2\pi) = sen\, x$$

$$\cos(x + 2\pi) = \cos x$$

It turns out that the form of each of these functions is constantly repeating itself. Moreover, our world is full of rhythms and periodic phenomena, such as day and night, the waves of the sea, the beating of the heart, the movement of the string of a guitar. Curiously, it was from the study of the latter that a real mathematical avalanche began".

The aim of this chapter is to analyse the knowledge acquired by students with respect to the contents of Trigonometry worked on during the year 2022 in Secondary Schools, upper cycle, in the city of Mar del Plata, Argentina.

In order to analyse what content they have seen or have stopped seeing, a survey was carried out among teachers in the last year of secondary school in different public and private schools in the city of Mar del Plata.

The results of this study show that the richness of trigonometry in secondary education is not being exploited, causing a gap in the horizontal knowledge of mathematics and hindering normal progress in the mathematics courses corresponding to the first years of university studies.

The teachers' responses lead us to conclude that there is a need to apply: pedagogical approaches that are adapted to the current context and to use resources that encourage deeper understanding.

Methodology

A survey was carried out among teachers in the sixth year of upper secondary school to find out which trigonometry concepts are worked on in their classes.

A Google form was used and disseminated by the Secretary of Educational Articulation of the Faculty of Exact and Natural Sciences of the UNMDP to upper secondary school teachers in the city of Mar del Plata (see ANNEX).

[to]The selected sample consisted of 55 teachers from public and public schools in the city of Mar del Plata, Buenos Aires, Argentina, who are in charge of the subject Mathematics in the 6th year of Secondary Education, in the Upper Secondary cycle. Of the teachers surveyed, 35 are from public schools and 20 from public schools.

Survey results

Data Analysis

From the data collected in the surveys, we observed that the orientations of the courses where teachers work are very varied.

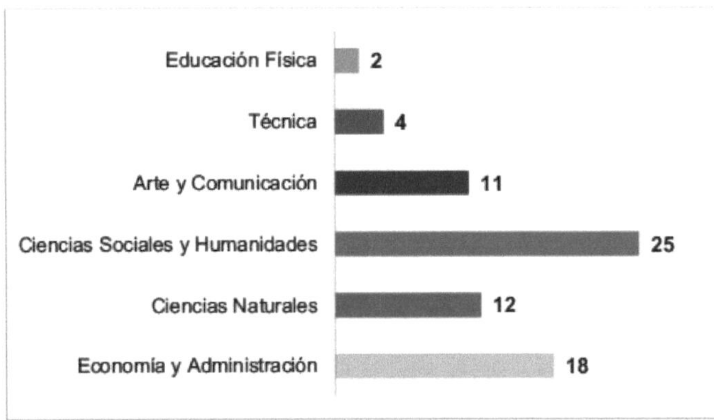

Physical Education
Technique
Art and Communication
Social Sciences and Humanities
Natural Sciences
Economics and Administration

Figure 1: What is the orientation of the course where you work?

Firstly, teachers were asked whether they taught Trigonometry in the sixth year of secondary school in the year 2022 (see Figure 2).

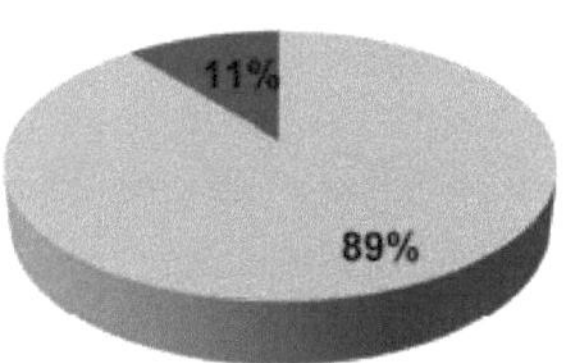

Figure 2: In 2022 did you teach trigonometry in sixth grade?

When asked about the reasons for not working on these topics, answers such as: "lack of time to retake unseen topics from the previous year of studies" or because "the students had worked partially on this topic in other years" appeared.

With regard to the question of whether your students had prior knowledge of Trigonometry.

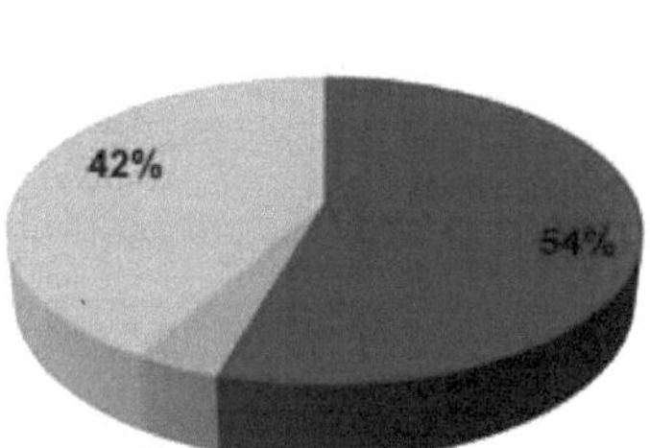

Figure 3: Did your students have prior knowledge of Trigonometry from previous years?

The 54% of teachers who responded that their students had prior knowledge of Trigonometry were asked specifically what this knowledge was (see Figure 4).

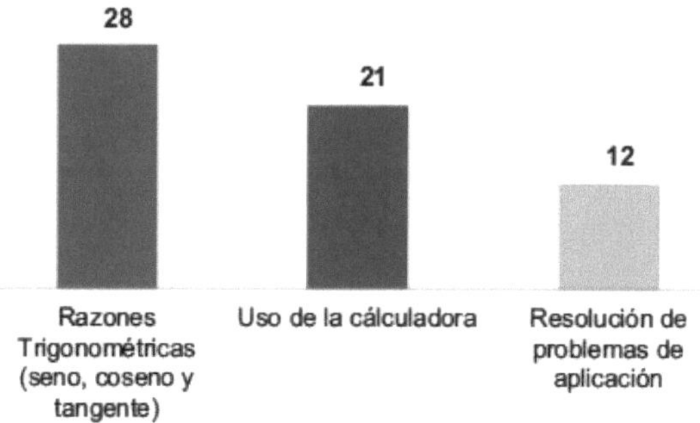

Trigonometric Ratios (sine, cosine and tangent)
Use of the calculator
Solving application problems

Figure 4: If your answer above was YES, what are these contents?

Most of the teachers responded that their students had previously worked with trigonometric ratios and that they had knowledge about the use of the calculator. To a lesser extent, their students used these contents for solving application problems.

Next, we inquired specifically about the teaching of Trigonometry in the sixth grade by asking the surveyed teachers to select those contents (see Figure 5). The topics of "Angle measurement systems: sexagesimal and circular" and "Graphs of trigonometric functions: general analysis" were the most frequently taught. Two teachers indicated that they worked with the sine and cosine theorem, however, this content does not correspond to the year under study.

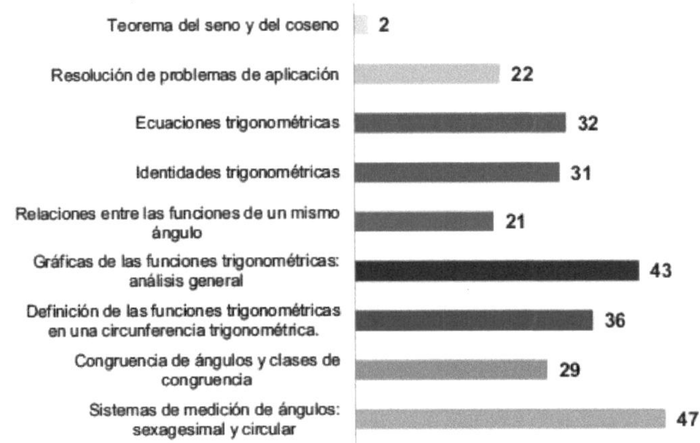

Sine and cosine theorem
Resolution of application problems
Trigonometric equations
Trigonometric identities
Relationships between functions of the same angle
Graphs of trigonometric functions: general analysis
Definition of the trigonometric functions on a trigonometric circle.
Congruence of angles and congruence classes
Angle measurement systems: sexagesimal and circular

Figure 5: Select the Trigonometry content you got to teach.

Finally, we were interested in analysing the strategies and resources used by teachers in the teaching of this content.

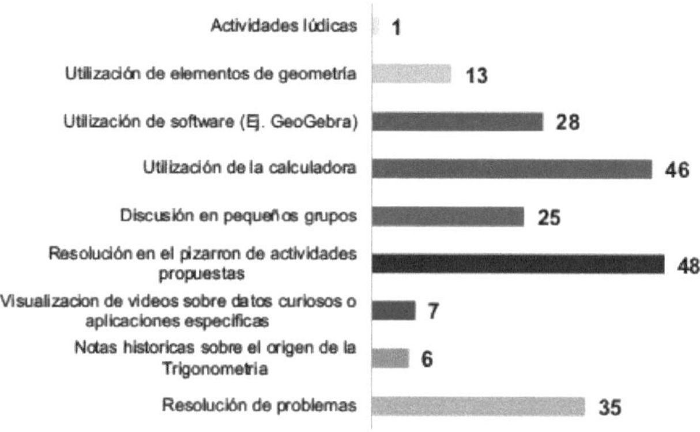

Playful activities
Use of geometry elements
Use of software (e.g. GeoGebra)
Using the calculator
Small group discussion
Resolution of proposed activities on the blackboard
Viewing videos on fun facts or specific applications
Historical notes on the origin of Trigonometry
Troubleshooting

Figure 6: Select the strategies and resources you used in the content dictation.

Some Reflections

The results obtained reflect that the time allocated and the depth of the treatment of the subject are far from adequate and raise questions about the preparation of students for the academic challenges they will face in the first years of university.

Moving forward positively requires:

- the application of technological resources.
- teachers who propose this type of knowledge with greater dynamism and who emphasise the formulation and resolution of problem situations applied in different areas of knowledge.
- students motivated and engaged in the process.

In order for students to fully understand a mathematical object, it is necessary for them to go through diverse experiences and practices. In this way, it will be the students who can organise and re-organise a network of relationships necessary to foster their autonomy in study. It is a complex challenge, but it is important that the student has control of his or her actions and can thus evaluate his or her own learning process.

To have mathematical ability implies to execute actions, to perform, to act appropriately and to reach a good end.

ANNEX

Trigonometry

The following survey is carried out by a research group in Mathematics Education of the Faculty of Exact and Natural Sciences of the National University of Mar del Plata. It is aimed at Mathematics teachers working in the sixth year of upper secondary school.

The answers are anonymous.

1. **The school where you work is managed**

Mara sOo Ln oval.

...private.

...publishes.

2. **JC What is the orientation of the course where you work?**

Select all the options that apply to you.

| | Economics and Administration

| | Natural Sciences

Humanities

| | Art and Communication

Technique

Others:

3. **cln 2022 did you teach trigonometry in sixth grade?**

Mark only Ln oval.

YES

No

4. **If your answer is NO, briefly describe the causes.**

5. **Did your students have knowledge of Trigonometry learnt in previous years?**

Mark only Ln oval.

YES

<; NO

CD NO SE

6. **If your answer above was YES, what are these contents?**

Select all the options that apply.

| | Trigonometric Ratios (sine, cosine and tangent)

Using the calculator

Solving application problems

Others:

If you taught Trigonometry in year 6 in 2022, answer the following questions.

7. **Select the contents of Trigonometry that you have taught**

Select all the options that apply.

| | Angle measurement systems: sexagesimal and circular Congruence of angles and congruence classes

Definition of the trigonometric functions on a trigonometric circle.

| | Graphs of trigonometric functions: domain, image: general analysis Relationships between functions of the same angle

| - Trigonometric identities

| | Trigonometric equations

Solving application problems

Others:

8. **Select the strategies and resources you used in dictating the content.**

Select all the options that apply.

Problem solving

Historical notes on the origin of trigonometry

| | viewing of videos on fun facts or specific applications Resolution on the blackboard of proposed activities

| | Small group discussion

Use of the calculator
Use of software (GeoGebra, Desmos, Photomath, etc.) Use of geometry elements (compass, ruler, protractor).
|_|_Others:

jThank you very much for your collaboration!

CHAPTER III

Valdez, Guillermo
Gamboni, Sandra

A didactic sequence related to the subject Trigonometry, for the sixth year of Secondary School.

In this chapter I am very pleased to share the design of a teaching proposal on the thematic unit: Trigonometry. It corresponds to the sixth year of the Escuela Polimodal Nr. 29, Centro Polivalente de Arte E.S.E.A. Nr. 1 of the city of Mar del Plata, province of Buenos Aires, Argentina, in classroom mode. It is a proposal in the Teaching Practices of one of my students of the Mathematics teaching program of the National University of Mar del Plata (currently, Professor: Sandra Gamboni), on the above-mentioned subject, which was original and very satisfactory in its application environment.

I can underline the originality of the teaching proposal, as I used new strategies for the students that contributed to an effective motivation, which was reflected in their productions in class.

It contains the following curricular elements: content, objectives, methodology and evaluation for the subject Trigonometry for the sixth year of secondary school.

We must consider the context for which the proposal is addressed, as a design cannot be adapted without considering the expectations, resources and social environment in which the students are immersed.

If such a context is not taken into account, it is likely that the expectations envisaged by the teacher will not be achieved, or that those that are achieved will fall short of the needs in each setting.

Esp. Guillermo Valdez

Characteristics of the School

The teaching practice took place in the Escuela Polimodal N°29, Centro Polivalente de Arte, E.S.E.A. N°1, located in a very old building in Diagonal Alberdi, corner Santa Fe. It is a public school, but it is attended by children from different social classes. It was created on May 2, 1974, together with four other establishments with the same characteristics in the Province of Buenos Aires. At that time, those who graduated had obtained the common secondary school diploma, as well as the National Folklore Teacher's degree. Later on, other educational offerings were added, such as Music, Instruments, Classical and Contemporary Dance, and the last one was Visual Arts. Graduates of this school have the possibility of entering, without the need to complete the Basic Training Cycle (Foba), a series of tertiary institutes in Mar del Plata, Buenos Aires and the respective career at the University of La Plata. At present, its enrolment is between 650 and 700 children, approximately. In reality, the total demand cannot be met due to the lack of space in the school. Children from other towns, such as Miramar, Santa Clara, Mar Chiquita, etc., attend the school, as the nearest school of the same kind is in Tandil.

It has a library and a video room, a photocopier and a buffet. In addition, there are many musical instruments, and rooms for practising dances. The co-operative plays an important role in the maintenance of everything, including the toilets and the leaks that often occur.

The class where this practice takes place is 6th grade, with a total of 29 students, 24 girls and 5 boys, who attend classical, folkloric and contemporary dance workshops. The classroom is located on the ground floor, next to the buffet, and faces Santa Fe street. There is a lot of neglect in terms of maintenance through the paint on the cracked walls and the wooden floors, with loose boards and a huge hole in one corner, where the children have to be careful not to step on or place the leg of a chair.

Pupils are regularly absent, to the point that all 29 pupils are never present. Sometimes it is because they are having a holiday, sometimes it is because they are rehearsing for a school event or party, and of course sometimes it is because of illness or personal issues. They also have the habit of being late by up to half an hour.

It is observed as a routine in this school, that the entrance is at 7:30 a.m. and between 7:40 and 7:50 a.m. the pupils, inside their respective classrooms, stand up to sing the song to the flag. After that the class begins.

Background

"Mathematics is not made to be observed, nor to see what others have done (and eventually get frustrated with it). No. Mathematics has to be made, transformed, improved, changed. And that can only be achieved by stimulating creativity".

Adrian Paenza

Trigonometry is a branch of mathematics that studies the relationship between the lengths of the sides of a triangle and the measure of its angles. Since ancient times it has been a useful tool to measure very large lengths, such as the radius of the Earth, the distance to the planets, the height of a tree or a mountain, etc.

In this sense, one of the main achievements is the one that allowed the cartographers to make the maps of the different countries.

But besides being used in cartography, astronomy and architecture, mainly for measurements, trigonometry is also of important application in other disciplines such as economics, meteorology, oceanography and biology, because trigonometric functions allow us to model situations that have periodic behaviour, such as the heartbeat.

And, of course, musicians also make use of these functions that give them the possibility to visualise the sound.

This topic provides an opportunity to address and integrate previous content acquired by students in previous years.

For its introduction, it is considered convenient to start from its application to carry out measurements, through an experience taken from real life, which will be carried out in the classroom by the students themselves with the guidance of the trainee teacher. According to a paragraph of the Diseno Curricular de la Provincia de Buenos Aires para 6° Ano, 2011:

"Although school mathematics *differs from scientific work, the style and characteristics of the work of the mathematical community can and should be experienced in the classroom. In this way, students will see mathematics as a possible task for all, as defined in the Basic Cycle of Secondary School*".

Objectives:

The objectives or expectations pursued are:

Teaching Objectives:

Present real-life problems where rigonometry can be applied.

Propose the study of the relationship between trigonometry and other disciplines, in order to appreciate its application in non-mathematical situations.

Encourage personal and group work, valuing individual and collective contributions to the construction of new mathematical content.

Promote respect for diversity of opinion.

Incorporate historical aspects of mathematics into the teaching of trigonometry.

Encourage students to evaluate their mathematical productions; make enquiries; defend; construct hypotheses.

Use algebraic language to symbolise, generalise and incorporate the trigonometric approach as another tool in problem solving.

Incorporate the use of new information and connectivity technologies (ntics) in the teaching of trigonometry.

Learning Objectives:

Recognise the usefulness of the application of trigonometry in real life.

Identify the sign of oriented angles and to which quadrant they belong.

Define the trigonometric functions and the trigonometric circumference.

Recognise the sign of trigonometric functions in the different quadrants.

To deduce that the functions depend only on the amplitude of the angle.

Construct the graphs of the trigonometric functions sine, cosine and tangent, from their respective representation on the unit circle.

To deduce relations between the trigonometric functions of the same angle.

Calculate the value of the trigonometric functions of notable angles in geometric form.

Perform the complete analysis of the trigonometric functions: sine, cosine and tangent.

Modelling everyday problems using trigonometric functions.

Classes

Class N^a^ 1: Presentation and Diagnosis

The teacher is introduced by means of a game where each student has to say his or her name out loud for a piece of candy. The evaluation guidelines are presented, emphasising that the mark is made up not only of the written evaluation, but also of the concept that includes: participation, the complete portfolio and teamwork.

Subsequently, activities are proposed in which they must apply previous contents that will be useful in the new subject: complementary and supplementary angles; triangles: classification according to their sides and according to their angles, similarity; ratios and proportions and the Pythagorean theorem.

Although the students were very enthusiastic to work in class, they did not remember most of the content involved in the diagnostic activities, so everything had to be explained to them.

Lecture N° 2: Introduction to trigonometry

Objectives:

J To arouse students' curiosity and interest in the study of trigonometry based on its usefulness and application in real life.

J Build an angle-measuring tool from items found in any household.

J Use trigonometry to measure a specific object.

Development: In previous classes we have begun to study the subject:

trigonometry. In this opportunity, we will return to this unit, but approaching it through one of its applications, such as the measurement of the height of an object. The teacher will begin by asking the meaning of the word trigonometry and then briefly explain to the students a historical fact of an application of trigonometry carried out by Eratosthenes to measure the circumference of the Earth.

tWhat does the word trigonometry mean?

Trigonometry is a part of mathematics that, generically, studies the relationship between the measure of the angles and the sides of a triangle. In fact, the word trigonometry itself has its origin in this fact: tri - means "three", gono - means "angle" and metria - means "measure".

Trigonometry already existed more than 3000 years ago, when the Babylonians and Egyptians used the angles of triangles to build pyramids and other complex architectural structures even for modern technology.

The stars in the sky inspired to further deepen trigonometry to discover their 'secrets' by creating star maps for calculating routes, predicting weather and space phenomena, clocks, calendars, etc.

In the last 100 years (approximately) one of the most important applications of trigonometry to mathematics is the study of wave and oscillatory phenomena, as well as periodical behaviour, closely related to the analytical properties of trigonometric functions.

Measuring the Earth

One of the major uses of trigonometry is to measure very large distances (or heights). Perhaps in this day and age, when we have so much technology at our disposal, we are not too impressed by what we will see now, but if we try to place ourselves some two hundred years before Christ, when there were no computers, no satellites, no mobile phones, no gps, not even the simplest calculator, then perhaps we will be able to arrive at the following conclusions

understand the importance of what this mathematician, geographer and astronomer Eratosthenes of Cyrene did.

Eratosthenes knew that, in Siena at noon on the summer solstice, if a stake was held vertically, it cast no shadow at all. That means that the sun's rays were perpendicular to the city. Our mathematician, moreover, believed that Siena and Alexandria were more or less at the same terrestrial longitude (i.e. on the same meridian) and that the Sun, being so far apart, would emit its rays in a parallel way between the two cities.

However, he was surprised when he measured the shadow of a stake (a stick) in Alexandria at the same time of the year, because the shadow cast had an angle of about 7.2°. What was happening? What was already known, that the Earth was not flat and that, when different points on the surface were subjected to the same light conditions, the shadows were different.

After observing this phenomenon and measuring the angle of 7.2° formed by the shadow in Alexandria, I searched the library for the distance between the two cities, which was estimated at 924 km. We see that all these data can be summarized in a very simple way: if the shadow cast in Alexandria had an angle of 7.2°, while in Siena there was no shadow at all, that meant that both cities formed a circular section of 7.2°, where the arc of circumference between the two cities was 924 km (see second part of the following figure).

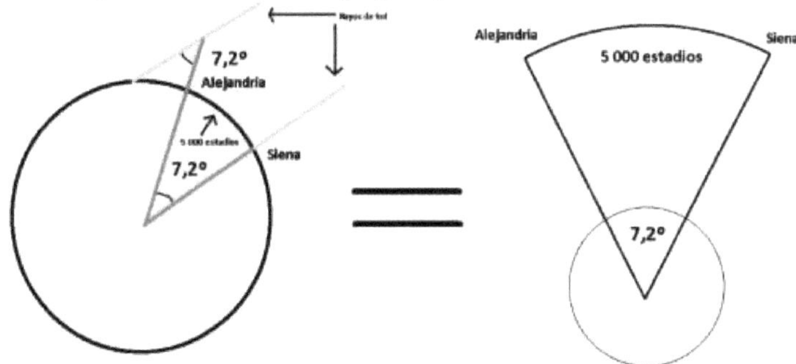

If we continue with the reasoning of Eratosthenes we can arrive at the following rule of three: if for an angle of 7.2° the arc is 924km, how long will the full 360° circumference be?

7, 2° 924 km.

360° X

If we multiply 360 by 924, and divide all that by 7.2 (so^ we would make a simple rule of three), we will see that then the circumference of the Earth, according to Eratosthenes, should measure 46 200 kilometres. Now, ^what is the real measurement? 40 075 km. That is, about 2 300 years ago, a man with a stick was able to measure the Earth with only 15% error.

In the age of Google Earth it may seem like too much of a mistake, but it's not bad for the first measurement of the Earth made more than two millennia ago.

In the age of telecommunications, you might not think it is a great achievement, but it is no mean feat that, in ancient times, when most people still thought the Earth was flat, a human being used little more than his wits to measure the circumference of the Earth. Eratosthenes' measurement opened the door to the impossible, for if someone had managed to measure the largest thing known to mankind with a stick, what could the scientists of the future not attempt?

The students will then be asked to construct a theodolite and then measure the height of a point on the blackboard using trigonometry. To do this, a copy of the instructions and the materials needed for the construction of the theodolite will be distributed to groups of 4 or 5 students.

Activity 1: Construction of a theodolite or clinometer

Materials:

- Large Conveyor
- Sorbet
- Glue
- Resistant thread (30 cm)
- Plumb line or hanging object.

Instructions:

- A small hole is made in the centre of the conveyor for the thread to pass through.
- The plumb line or object is attached to one end of the thread, and the other end is attached to the protractor.
- Glue the sorbet to the protractor so that it lies along the 0° and 180°.

It should look like the figure:

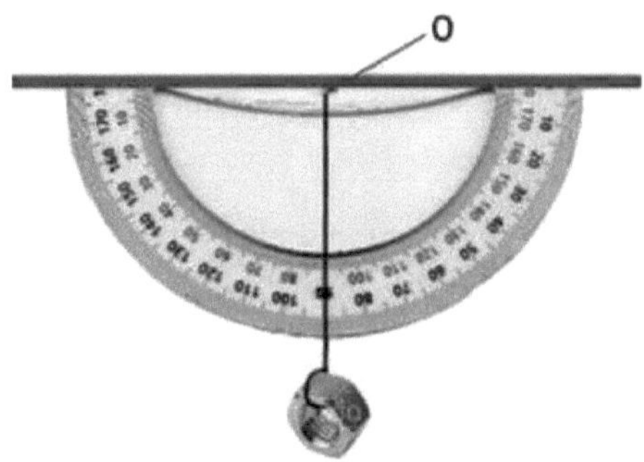

Activity 2: measuring a height

Measure how high above the ground a point marked on the blackboard is, using the theodolite. The measurement will be done 3 times, and then the average will be taken. The work will be done by the students in teams of 4.

The point on the blackboard is then measured with a tape measure to compare with the results obtained by the students. The team that comes closest to the actual measurement wins the competition.

To measure the angle you have to take into account, ^what is the angle to be measured?

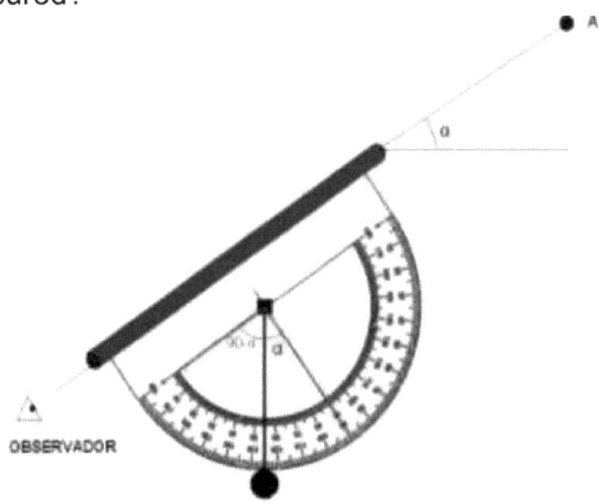

As can be seen in the figure, the angle measured with the theodolite is actually 90° - a.

For the measurement the observer has to stand at a known distance from the blackboard wall. From there he will measure the angle with the theodolite. The data collected will be recorded in the following table.

Measurement	Angulo	Distance	Observer height	Height sought	Total height

1					
2					
3					
Average					

The outline of the problem is shown in the figure below:

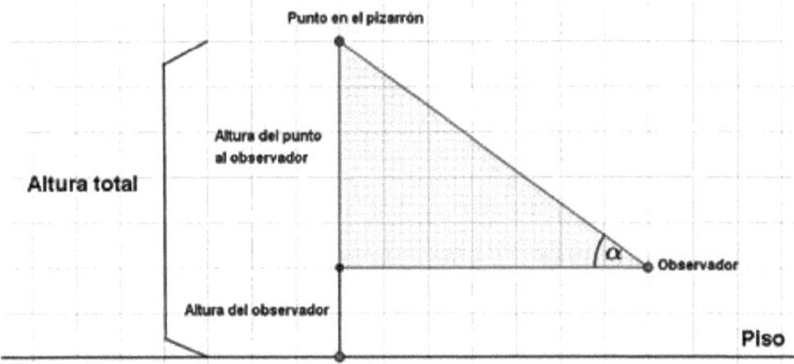

Note:

The pupils were very attentive to the historical introduction and were very interested and somewhat surprised.

The construction of the theodolite was a great motivation for the students. This kind of activity with concrete material was very enjoyable for them. They were then able to use it to measure the angle they needed to calculate the height of a point on the blackboard.

At the end of the class the general comment from the students was: "We never had such a fun maths class".

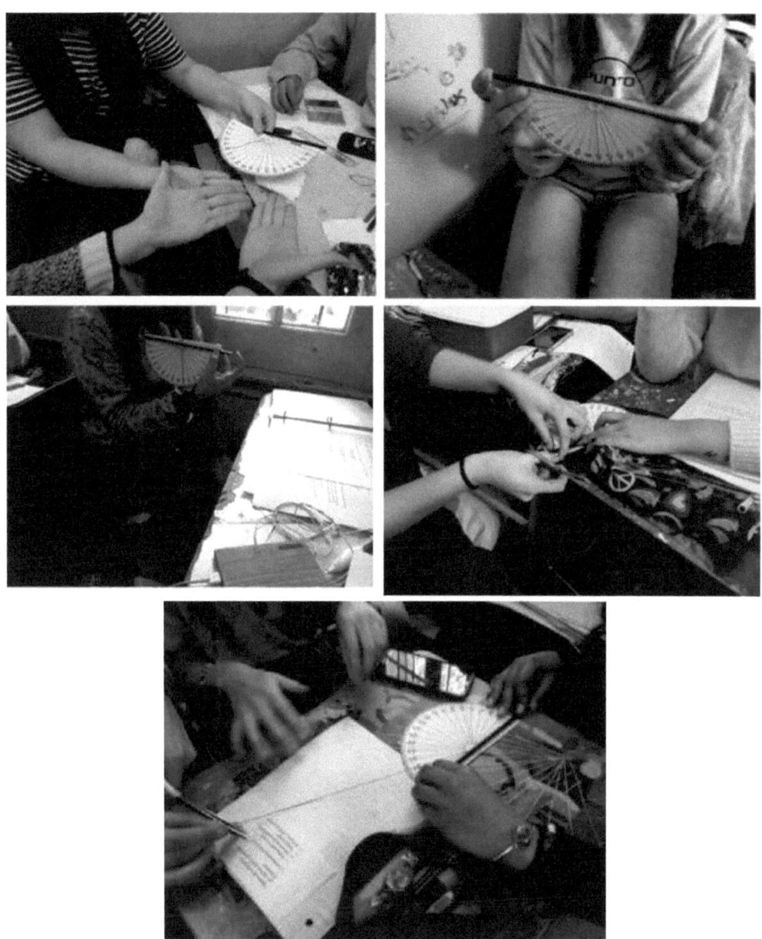

Lecture N°3: Trigonometric ratios depend only on the angle

They were given a problem in which they were asked to build a ramp for which there were two possible designs from which they had to choose the one with the least angle of inclination to the ground. To carry out the activity, the students were given the triangles of the printed designs so that they could cut them out and superimpose them and thus recognise that in reality, although the triangles were of different sizes, the angles measured the same:

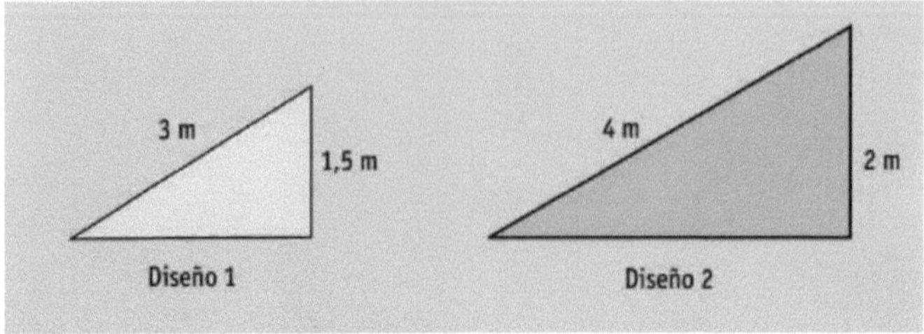

When posing the following problem of the choice of the ramp between two designs, a small debate arose between some students who gave their opinion whether the answer was design 1 or design 2. With the proposal to cut out and superimpose the triangles, they were able to observe with concrete material that the respective angles of the two triangles have the same width. It was concluded that the triangles were similar.

Lecture N°4: Trigonometric chain (play activity)

A playful activity is proposed that consists of a chain of cards in which, on one side, there is a triangle in which a certain piece of information has to be calculated, and on the other side, the result of another card.

The aim is to form a chain of cards, one of which has the result of the other. The game consists of two chains. Therefore, the students will form two groups. In each group all the cards will be dealt and they will have to put the chain together. The first group that manages to put it together will be the winner.

While the children were enthusiastic, there were a few hiccups. The first hitch they encountered was the absence of scientific calculators, but they were shown that they had one on their mobile phones, and were taught how to use them, which took a considerable time. However, it was useful because now the children were familiar with the calculator on their mobile phones, which in the past we had tried to do, but they did not pay much attention to it. Instead, in the middle of the game, as they wanted to win, they were more eager to learn.

Lecture N° 5: Trigonometric Ratios and Ratios in Approximate Ratios

To introduce the trigonometric ratios we start with a problem where we are asked to find the measures of the legs, and the amplitude of the acute angles of a right triangle, of which we know how long the secant of one of its acute angles and its hypotenuse are.

In this way, it is explained to the students that these trigonometric ratios are not found on the calculator and therefore, in order to calculate them, it is necessary to transform them into their corresponding ratio. That is to say, for example, if we are dealing with the secant, as in the case of the problem posed, we must use the cosine, which is its netproca to enter it in the calculator.

Lecture N°6: Oriented angles

Angles are defined as positive when they are generated by counterclockwise rotation, and negative when they rotate clockwise; furthermore, they have an initial side and a terminal side.

Given a system of Cartesian axes in the plane, we consider an angle oriented in the plane as follows:

Its **vertex** is the origin of coordinates.

It is generated by the rotation of a ray with origin in (0;0). The initial position of the ray coincides with the positive semi-axis of the x-axis (**initial side**) and it rotates keeping its origin fixed until it reaches a position that marks its **terminal side**.

Its sign is determined by the direction of rotation of the initial side, according to the convention mentioned above.

To indicate its location, we locate the terminal side in one of the four quadrants into which the plane is divided by considering the Cartesian axes.

Lecture N°7: Trigonometric Circumference

To define the trigonometric circle we start from a question: ^Can we calculate the sine and cosine of angles whose amplitude is greater than 90°? If yes, ^How can this be done?

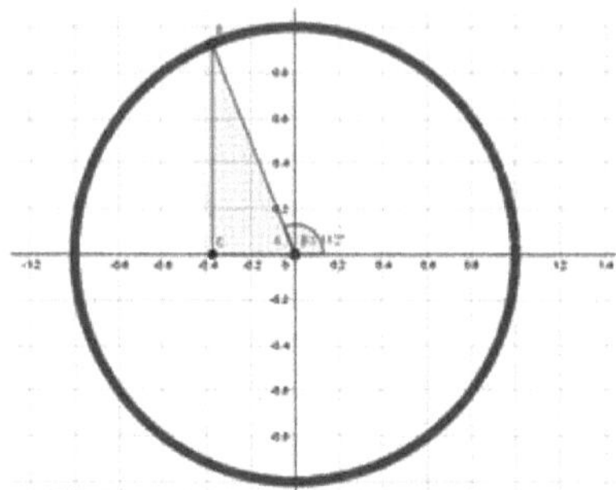

To answer these questions, a right triangle with hypotenuse of measure one is taken. With the help of the GeoGebra software, the students are given an activity in which they have to create an animation in this application. This animation will rotate the hypotenuse and, as if it were a compass, it will draw the trigonometric circle of radius 1.

From this circumference the trigonometric functions can be defined. This activity with the application of GeoGebra produced a lot of interest on the part of the students to learn how to use this program.

Class N°8: Practical Work to be handed in

A homework assignment is proposed to be done in class and then handed in, with the aim of reinforcing the concepts of solving right triangles, and identifying the hypotenuse and the opposite and adjacent legs.

The following errors were observed in the correction of this activity:

- They do not identify the hypotenuse, nor do they distinguish the opposite from the adjacent leg.
- Notation errors such as *a;b o a - b* to indicate a side *ab*
- They do not recognise that sine, cosine or tangent must be applied to an angle and write: *sin* = -

It is considered that the activity then fulfilled the objective of being able to detect these errors in order to clarify them with the students.

Lecture N°9: Notable angles

To approach this topic, the students are given a graduated circumference as shown in the figure, with the aim of having them write in each angle measure that is in the sexagesimal system, its equivalent in the circular system. In this way, it is intended that the students have well incorporated this system of measurement so that they can use it in the graphs of the trigonometric functions.

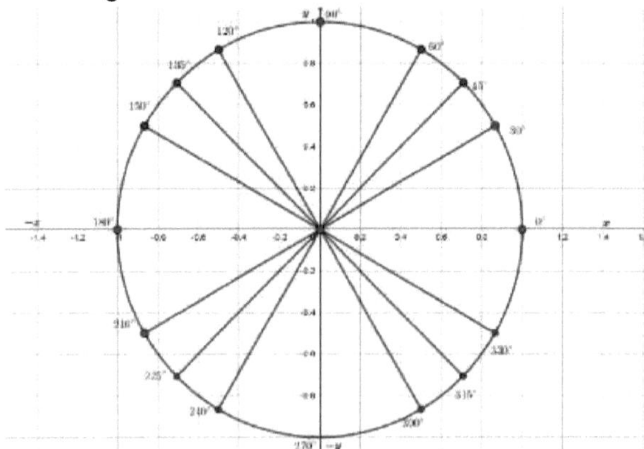

They must then complete a table with the trigonometric functions of the notable angles:

	0	π/6	π/4	π/3	π/2
Seno	0	$\frac{1}{2}$	$\frac{\sqrt{2}}{2}$	$\frac{\sqrt{3}}{2}$	1
Coseno	1	$\frac{\sqrt{3}}{2}$	$\frac{\sqrt{2}}{2}$	$\frac{1}{2}$	0
Tangente	0	$\frac{\sqrt{3}}{3}$	1	$\sqrt{3}$	$\nexists$

The problems encountered in learning this subject are undoubtedly that it integrates many other previous contents that should have been seen in other years, such as isosceles triangles, Pythagoras' Theorem, clearing quadratic equations that involve the use of absolute value, etc. and the students do not remember them.

Lecture N°10: Analysis of trigonometric functions

To introduce the analysis of trigonometric functions, their real-life applications are discussed, with emphasis on their use in modelling periodic situations such as the heartbeat and the sound wave, in order to motivate students to study these functions. An experiment is carried out with a home oscilloscope, and then they are presented with a virtual oscilloscope and a scenario in GeoGebra showing the sound wave in which the amplitude and frequency can be varied.

Then it is explained how to graph the functions starting with the sine and cosine, for which they should use the graduated circle with which they worked in the previous class.

Subsequently, the analysis of these functions is carried out, detailing: domain, image, rafc;es, order to the origin, sets of positivity and negativity; intervals of growth and decrease. In addition, students are asked what happens when we want to find the sine or cosine of an angle that measures more than one turn, with the aim of observing the periodicity of the function, since it takes the same values again.

The greatest difficulty the students had was in the intervals of growth and decrease and in the sets of positivity and negativity, because it was difficult for them to see that they had to give it on the x-axis.

The other obstacle was in graphing the sine function, because they were not yet afraid of passing the angles to the circular system on the trigonometric circle that they had been given in the previous class, and it was difficult for them to mark the points on the Cartesian axis system.

Lecture N°11: Analysis of the tangent function

Continuing with the study of trigonometric functions, it is the turn of the tangent. In order for the students to be able to carry out the analysis, they are given a graph of this function, from which they have to

determine its domain, image, rafc;es, order to the origin, positivity and negativity, growth and decrease.
The main problem, in this case, is that the tangent is not defined for all the real numbers, and vertical asymptotes are observed in the values where it does not exist. In spite of the fact that, in the previous year, the students have already seen what an asymptote is in other functions, such as the exponential or the logarithmic, they have some difficulties in understanding this concept.
This issue is also closely linked to the fact that in mathematics **it is not possible to divide by zero.** Precisely, the real numbers in which the tangent is not defined are those in which we would be left with a division by zero. In general, students seem to confuse division by zero with division by one, since, when they encounter an expression, for example, if we have:

$$tg(90°) = \frac{sen(90°)}{\cos(90°)} = \frac{1}{0} = \nexists$$

While we know that there is no value that fulfils this, students often answer that the result is one.

Lesson N°12: Building a machete

By means of a heuristic discussion with the students, we try to establish which are the main concepts they need to know for the evaluation, while at the same time a synthesis of these concepts is written on the blackboard.
It is interesting that, within the revision class, the construction of the machete or summary is included, given that one of the most common disadvantages seen when studying and preparing for a mathematics exam is that students definitely do not know how to study mathematics, as it is not a matter of reading and underlining the main ideas as is the case with text-heavy subjects.
students definitely do not know how to study mathematics, since it is not a matter of reading and underlining the main ideas as is the case with text-heavy subjects. For this reason, it is helpful to guide them in this matter, showing them which concepts they need to handle in order to solve the exam.

Class 13: Review activities

In the class prior to the exam, students are given integration activities that involve everything they have seen in the unit to be assessed. For this topic, they will have to solve a problem with a right triangle, where trigonometry needs to be used, such as calculating the length of a ladder leaning against a wall, knowing the angle it forms with the floor or the angle it forms with the wall itself, and the height at which it is leaning.

Also, students will need to know the analysis of the trigonometric functions that were previously studied, because they may encounter activities in which they have to determine the domain, the image, the ratios, and everything that is related to the study of these functions.

Other activities that will form part of the written assessment are those that have to do with trigonometric identities, such as finding all the trigonometric functions from a given one without calculating the angle. In this case they will need to use the relation that states that the tangent is the quotient of the sine and the cosine, and the Pythagorean identity.

The only subject in which the students do not have any problems is the analysis of functions. In the rest of the subjects, there are many difficulties.

Class 14: Written evaluation

Written assessment is carried out. The elements that students are allowed to have in the exam are:

J Trigonometric function graphs

J Summary sheet of formulas written in class: definition of trigonometric ratios and identities.

J Calculator (Since many of the students do not have a scientific calculator, the values of the sine, cosine and tangent of the angle in question are presented in the corresponding activity so that they have to decide which one to use in the given situation).

25 students attended the evaluation, 4 of them were absent. Of those who took the exam, four handed in a blank, three of them have already taken the subject, but the rest completed all the items.

The result was that 69 % of the students passed the exam, 24 % failed, while 7 % were absent.

The graph below illustrates the proportion of pass, fail and absent students:

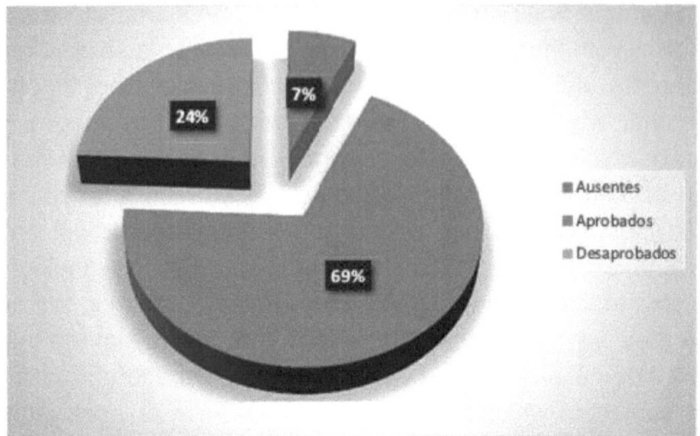

A model of the summative assessment taken is shown below, followed by a description of the analysis of the most common errors made by the students:

1) Determine whether the following statements are true or false. If false, justify:

a) In the third quadrant the cosine is positive.

b) The tangent is increasing throughout its domain.

c) There is an angle whose cosine is -2.

d) is a zero of the tangent function.

2) On each of the following circles is indicated the sign of the trigonometric ratios of a, according to the quadrant in which it lies. Sign the circle that corresponds to the **cosine**.

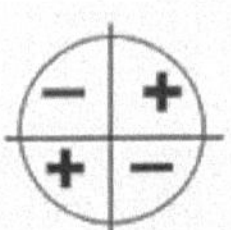

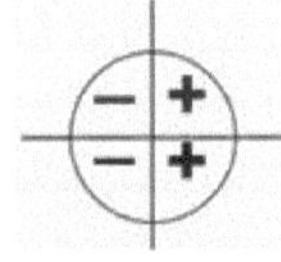

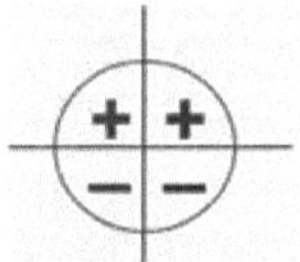

3) a) Write in circular system: 260° and 120°.

b) Indicate to which quadrants the above angles belong.

4) Determine the other trigonometric functions of *a* knowing that:

$\text{sen}(\alpha) = \frac{\sqrt{2}}{2}$ and a belongs to the second quadrant.

5) Find the distance from a ladder to the wall, taking into account the data in the figure:

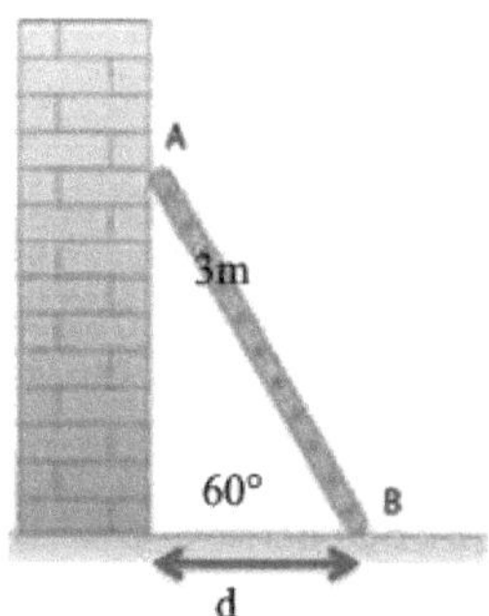

Additional data: $sen(60°) = \frac{\sqrt{3}}{2}$; $\cos(60°) = \frac{1}{2}$; $tg(60°) = \sqrt{3}$

Solution: 1)

a) False. The cosine is positive in the first and fourth quadrants.
b) True.
c) False. The cosine image lies between -1 and 1.
d) False. The tangent does not exist at that value.
2) The second is the cosine.
3) a)

$$260° = \frac{260°.\pi}{180°} = \frac{13\pi}{9}$$

$$120° = \frac{120°.\pi}{180°} = \frac{2\pi}{3}$$

b) The angle of 260° belongs to the third quadrant and the angle of 120° to the second quadrant.

4) Since we know the value of sine, we can find the value of cosine using the Pythagorean identity

$$\left(\frac{\sqrt{2}}{2}\right)^2 + \cos^2(\alpha) = 1 \rightarrow \cos^2(\alpha) = 1 - \frac{1}{2} \rightarrow |\cos(\alpha)| = \frac{\sqrt{2}}{2}$$

As a belongs to the second quadrant there the cosine is negative, so we are left with:

$$\cos(\alpha) = -\frac{\sqrt{2}}{2}$$

We find the tangent:

$$tg(\alpha) = \frac{sen(\alpha)}{\cos(\alpha)} = \frac{\sqrt{2}}{2} . \left(-\frac{2}{\sqrt{2}}\right) = -1$$

Finally, we find the reciprocals:

$$\sec(\alpha) = \frac{1}{\cos(\alpha)} = -\frac{2}{\sqrt{2}} = -\sqrt{2}$$

$$cosec(\alpha) = \frac{1}{sen(\alpha)} = \sqrt{2}; \quad cotg(\alpha) = \frac{1}{tg(\alpha)} = -1$$

The distance from the ladder to the wall would be the cathetus adjacent to the 60° angle, and we have the length of the ladder as the hypotenuse, so we use the cosine:

$$\cos(60°) = \frac{d}{3} \to \frac{1}{2} = \frac{d}{3} \quad \to d = \frac{3}{2} = 1{,}5\ m$$

Among the most striking errors we found that, for example, some answered wrongly to item 1) a) which asked them to say whether the statement that in the third quadrant the cosine is positive was true or false, while they chose the correct option in item 2 where they had to select the graph representing the signs of the cosine. In short, it was the same question, but graphically. They only had to associate the items.

Another significant error can be seen in the misuse of the calculator, and simply in the accounts, since in order to find the values of trigonometric functions or angles, the use of the calculator was reduced as much as possible because there are many students who are not afraid of having access to a scientific calculator.

Copying errors were also observed in the sequence of the steps that the students themselves proposed to solve the exam activities, especially in item 4 where they had to find all the trigonometric functions from a given one. Some students, from one step to the next, forgot to copy a sign or changed the denominator. These mistakes are often attributed to distractions.

In item 3, the problem was in determining to which quadrant the given angles belonged. Some students omitted this item altogether.

As for item 5 in which a problem was posed where there was clearly a right triangle and it was only necessary to determine which trigonometric reason should be used, the error was precisely in this election, confusing the cosine with the sine.

According to the percentages of correct resolution of each Rem, it is observed that the easiest one was Rem 2, in which they had to indicate which was the graph corresponding to the signs of sine or cosine, depending on the subject they had been given; and the most difficult one was Rem 4, where they had to find all the trigonometric functions starting from a given one. This result coincided with the prediction of the trainee teacher, and precisely the score given to each of these Rems reflects this prediction, since Rem 2 was the one with the lowest score (1 point), and Rem 4 the one with the highest score (3).

Class N°15: Levelling

The evaluations taken in the previous class are returned. In view of the fact that the absentees have to take the test, the idea of solving the previous evaluation on the blackboard is abandoned. Instead, you clarify individually with the children the mistakes they made as they receive their tests.

Personal appreciation of the trainee teacher:

The experience of the Internship was very rich in all aspects. There were very beautiful and rewarding moments and others not so much. To be more specific in my appreciation I will detail what I found: positive, interesting and negative (PIN).

- Positives:
  - The opportunity to stand in front of a class with a blackboard behind me and chalk in my hand, because such a situation fascinates me, and I really enjoyed all those moments when I was interacting with the students, explaining and guiding them to reason, deduce and think for themselves.
  - The Wednesday classes where the students were always willing and more awake than the Tuesday classes. It was incredible to be able to appreciate this difference because from one day to the next they seemed like different people.
  - The relationship I established with the students in general and especially with some of them who participated more in the classes or showed more interest in learning, even asking questions outside the subject that was being given, encouraged me to investigate and learn more.
  - The performance of the students in the exam, having passed almost 70% of the course, and with very good marks.
  - The support, in general, of the head teacher, although on some occasions she mimicked the natural interruptions of the school.
- What's interesting:

o The first class I gave, unique and unrepeatable, when we set up the theodolite and they were amazed because they had never thought that a mathematics class could be done in another dimension outside the paper.

o I realised after the end of the classes and when I started to put together the portfolio for the practicals how useful the sociogram was and how well it reflected the characteristics of the children and the group.

o One of the boys who had the hardest time understanding the subject was able to get an A in the exam, but the most interesting thing was to see his happiness. And in every class that he managed to understand something, after several times of explaining it, he always said to me with a smile: "I get it, I get it", and I think he did it not only because he was happy but so that I would be happy too.

- The negative:

o The teachers' stoppages and all the interruptions in the classes which meant that the development of the subject took longer than necessary and that, as there was no continuity, the students were unable to link the concepts as expected. In addition, content such as trigonometric equations and identities could not be covered.

o The last class where some of them expressed their dissatisfaction, verbally or through the test I asked them to take, with my way of dealing with or explaining the subject. Although they perceived that the discontinuity in the classes had been a fundamental obstacle in the learning of the contents, some were not satisfied with my work. At the beginning I was a bit disappointed, but as the days went by I understood that it was logical for them to see it this way. They were used to solving exercises in a rather mechanical way and not so much to thinking, reasoning and deducing by their own means. In order to deduce new concepts, some previous knowledge was required, which they did not fully master.

In summary, and taking stock, the experience turned out to be more positive and interesting than negative. And even the negative part helps us because it shows us where we have to improve in order to achieve a richer experience in the future. Anyway, every school, every course, every moment is unique.

Sandra Gamboni

CHAPTER IV

Valdez, Guillermo
Vecino, Susana
D^az Lanzoni, Evelyn
Tissoni, Gaston

Creating an alternative space: GeoGebra, Complex Numbers and Trigonometry

Gonzalez et al. (2017), states that "in the traditional teaching of Trigonometry, the figures drawn by the teacher on the blackboard, in addition to being static and rigid, can be very different from what the teacher wants to represent" (p. 401). For this reason, he suggests the use of various software and in-formative programmes such as GeoGebra, where the student can, from a construction, alter the objects while preserving the original characteristics.

This chapter is written primarily to provide secondary school and/or first-year university mathematics teachers with a set of organised activities for them to experiment with technology-mediated mathematical production work.

Because of the type of content that is required as basic, the activities can be adapted for the sixth year of secondary school and for the first year of university.

On 27 September 2023, two advanced students of Mathematics under the supervision of the teachers in charge of the subjects Algebra and Linear Algebra I, carried out an intervention using two GeoGebra applications with first year students of Mathematics, Physics, Biochemistry and Chemistry of the Faculty of Exact and Natural Sciences of the National University of Mar del Plata.

GeoGebra is a dynamic geometry software, which is freely accessible and multiplatform. As its name indicates, it is not only a software that allows the work with geometrical contents (Geo) but also allows the interaction with algebraic contents (Gebra), analytical and statistical. That is to say, GeoGebra allows the double perception of objects. Each object has two representations, one in the graphical view (Geometry) and one in the algebraic view (Algebra).

The idea of this meeting was to present to the students two applications in GeoGebra (usable both on PC and mobile phone), designed by Prof. Jose Campos, teacher of our academic unit. These applications allow the students to graph complex numbers, perform product and quotient operations, calculate modulus, argument, and the nth ratios of a complex and represent their respective graphs. *Trigonometry"* plays a fundamental role in the development of all these concepts.

More than 40 students from the aforementioned degree courses attended, who were taking the subjects Algebra and Linear Algebra I. The minimum contents of these subjects are: Natural Numbers, Complex Numbers, Polynomials, Matrices and Determinants, Systems of Linear Equations and Vector Spaces.

It should be noted that their participation was entirely voluntary, as the meeting did not take place in a class timetable corresponding to the subjects, but consisted of an extra-curricular workshop activity.

At the beginning of the activity, students were given a series of activities to solve using the applications (see below).

COMPLEX NUMBERS

1) Using one of the GeoGebra Applets presented, complete the following table.

$Re(z_1)$	$Im(z_1)$	z_1	$Re(z_2)$	$Im(z_2)$	z_2	$\lvert z_1\rvert$	$\lvert z_2\rvert$	$\lvert z_1 . z_2\rvert$	$Arg(z_1)$	$Arg(z_2)$	$Arg(z_1 . z_2)$
1	-2		-2	1							
		1+i			1-i						
2	1				2+2i						

Then answer the following questions:

a) What is the relationship between $|z_1|, |z_2|$ y $|z_1 . z_2|$? ¿ How would you express it in your words? ^and algebraically?

b) What is the relationship between $Arg(z_1), Arg(z_2)$ y $Arg(z_1 . z_2)$? ¿ How would you express it in your words? ^and algebraically?

c) ^ What relationship can be established between $|z_1|, |z_2|$ y $\left|\frac{z_1}{z_2}\right|$, with $z_2 \neq 0$? ¿Y between $Arg(z_1)$, $Arg(z_2)$ y $Arg\left(\frac{z_1}{z_2}\right)$, con $z_2 \neq 0$?

2) Using the GeoGebra Applet that involves ennezimal roots, work with the following complexes:

a) $2i$, con $n = 5$.

b) $2 - 2i$ con $n = 3$.

Find the nth roots of the given complex, in polar form.

Relate the graphs to the value of n. How do the moduli of the nth roots relate to the given complex? ^How do the arguments vary?

Once you have completed the activities, we ask you to answer the following questions about GeoGebra Applets:

With the use of the Applet, did you find these activities difficult?

Would you use any of the applets in the practice activities?

Do you think applets are a good tool to work with complex numbers? ^By What?

Applets:

Taking the first section of the activity as an example, a brief explanation of how the GeoGebra applications work was given, the access link to which can be found in the respective QR.

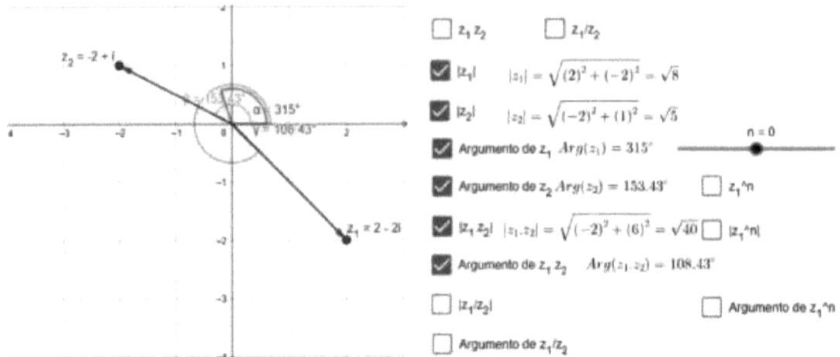

It was agreed with the students that in a period of approximately twenty minutes they would solve the remaining items using the resource presented, which can be accessed by scanning a QR code (provided in the photocopies of the activities) with their mobile phones.

Once the proposed activities had been solved, there was a sharing session on the blackboard, with everyone carrying out the constructions together and then giving a graphic interpretation of the results.

The students were very interested in the visualisation of the graphs.

After approximately 15 minutes, another sharing took place, with the same modality for the exercises referring to the nth rafces.

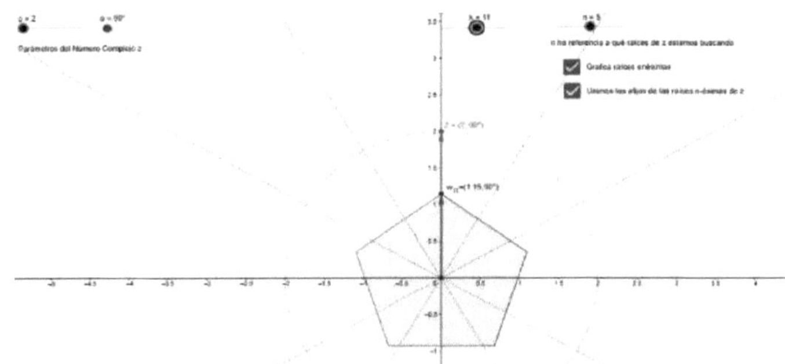

Finally, those present were asked to answer a short survey: *"With the use of the Applet,*

- *Did you find these activities difficult?*
- *Would you use any of the applications in the practice activities?*
- *c Do you think these applications are a good tool to work on complex numbers? ^Why?"*

From this survey, some important reflections emerged from the students, which will be highlighted below:

First of all, most of the students considered that the GeoGebra resource allowed them to quickly verify results and speed up calculations .

As we can see in the following case, a student expresses that he will start using the applications to check that the activities proposed by the practical part of the subject are well done.

①¿TE RESULTARON DIFÍCILES ESTAS ACTIVIDADES?
NO.
② ¿UTILIZARÍAS ESTE APPLETS EN LA ACTIVIDADES
③ DE LA PRÁCTICA? Sí, ME PARECE UNA HERRAMIENTA
④ MUY PRÁCTICA, Y LA VOY A EMPEZAR A USAR PARA
COMPROBAR QUE ESTÁN BIEN HECHOS LOS EJERCICIOS

At the same time, another student says that he finds it difficult to cope with technology, but believes that working on a computer would be easier for him.

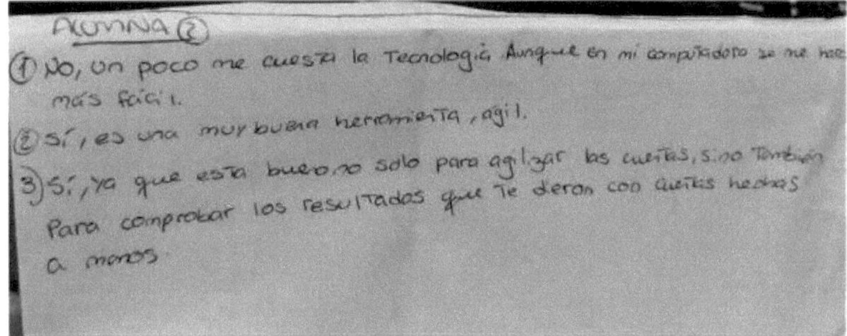
ALUMNA ②
① No, un poco me cuesta la Tecnología. Aunque en mi computadora se me hace más fácil.
② Sí, es una muy buena herramienta, ágil.
3) Sí, ya que está bueno no solo para agilizar las cuentas, sino también para comprobar los resultados que te dieron con cuentas hechas a manos.

Another important conclusion is that they not only use it to check their answers and correct possible mistakes, but also to observe graphically what they have worked on.

¿Utilizarías en las actividades de la práctica alguno de los applets
Sí los utilizaría.
¿Crees que los applets son una buena herramienta para trabajar los números complejos? ¿Por qué?
Sí creo que son una buena herramienta porque te permite ver gráficamente lo que trabajamos de manera analítica y corregir mis propios errores.

What do the advanced teacher training students who participated in the experience think?

In general terms, we consider that the intervention in the classroom to present the applications was really positive. The students were interested and participated well in solving the activities. In addition, the survey we carried out allowed us to know their opinion about the possible use of these tools and, at the same time, they presented us with some criticisms or observations, which we consider as suggestions to improve their functioning.

What did the responsible teachers think?

For the teacher who brings to the classroom a type of activity such as those proposed here, it implies a work of accompaniment to their students in different processes.

For example:

- in the processes of appropriation of the technological knowledge necessary for working with GeoGebra,
- in the process of producing new ideas,
- in the processes of reconfiguration of knowledge already learnt at other educational levels (in this case in particular previous concepts of trigonometry),
- in the processes of conjecture production and validation.

The proposed work proposal seeks to offer an alternative space for mathematical work in the classroom, which involves adaptation strategies for those who learn and also for those who teach, and at the same time a reinforcement of the basic concepts of trigonometry that are applied in this subject.

Bibliography

- Amster, P., Pezzatti,L., Roxana Abalsamo and others. Activados 5 Matematica. Editorial Puerto de Palos. Buenos Aires, Argentina, 2013.
- Benavente,M., Vivera,C. Introduction to Mathematics, Modulo de Ingreso Universidad Nacional de Mar del Plata, 2015.
- Bracchi,C., Paulozzo,M., Diseno Curricular para la Educacion Secundaria, 6to. Ano, Matematica, Ciclo Superior, Direction General de Cultura y Educacion de la Provincia de Buenos Aires, 2011.
- Cantoral, R., Montiel, G., & Reyes, D. (2015). Analisis del discurso Matematico Escolar en los libros de texto, una mirada desde la Teona Socioepistemologica. Avances de investigacion en educacion matematica, 5(8), 9-28.
- de Guzman, M., Colera, J. and Salvador, A. (1988). Matematicas. Bachillerato 3. Grupo Anaya S.A. Madrid, Spain.
- Direction General de Cultura y Educacion (2011). *Diseno Curricular para la educacion Secundaria. Matematica Ciclo Superior. Sexto ano.* Buenos Aires: Autor. Recuperado de: Disenos curriculares | abc
- Gonzalez, M., Matilla, J., and Rosales, F. (2017). Potencialidades del software Geogebra en la ensenanza de la matematica estudio de caso de su aplicacion en la trigonometria. Roca: Revista Cienrifico Educaciones de la provincia de Granma, 13(4), 401-415. Retrieved from https://dialnet.unirioja.es/servlet/articulo?codigo=6759725Herrera (2013)
- Leon, C. (2017). Juan Cortazar y su contribution a la formation matematica espanola en el siglo XIX (Doctoral thesis). University of Cordoba, Cordoba, Spain.
- Ministerio de Educacion, Cultura, Ciencia y Tecnolog^a (2018) Argentina en PISA 2018. Informe de Resultados. Retrieved from: https://www.argentina.gob.ar/sites/default/files/argentina_en_pisa_2018_infor me_de_resultados.pdf
- Ministry of Education Argentina. Secretaria de Evaluacion e Information Educativa. APRENDER 2022 results. Retrieved from: https://www.argentina.gob.ar/sites/default/files/2023/06/_resultados_a_nivel_ national_learning_2022_at_secondary_level_-secretary_for_educational_information_and_evaluation.pptx_1.pdf
- Robert, A., & Rogalski, J. (2002). Le systeme complexe et coherent des pratiques des enseignants de mathematiques: Une double approche. Canadian Journal of Science, Mathematics and Technology

Education, 2(4), 505-525.
• Paenza,A., Matematica...^estas ahi?, la vuelta al mundo en 34 problemas y
8 stories. Grupo Editorial Siglo Veintiuno, 2011.
- Zill, D., Dewar,J.. Algebra, trigonometry and analytic geometry. Third Edition. Mc Graw Hill, Mexico, 2012.

Internet resources:

• https://anagarciaazcarate.wordpress.com/2018/04/10/dos-cadenas-trigonometric/
• http://didacticaenlaciencia.blogspot.com/2008/06/qu-vamos-hacer-vamos- build-un.html
• http://servicios.abc.gov.ar/lainstitucion/revistacomponents/revista/archivos/ textos-escolares2007/CM-EN5-1PA/filesfordownload/CM_EN5_1PA_u4.pdf
• http://servicios.abc.gov.ar/lainstitucion/revistacomponents/revista/archivos/ textos-escolares2007/CM-EN5-1PA/filesfordownload/CM_EN5_1PA_u5.pdf
• http://bdigital.unal.edu.cc/49554/1/11706629.2015.pdf.pdf
• http://ulum.es/historia-de-las-mediciones-i-eratostenes-el-hombre-que- midio-la-terra/
- https://todaslascosasdeanthony.com/2012/07/02/como-eratostenes-midio- the-circumference-of-the-earth-2-thousand-years-ago/
• http://recursostic.educacion.es/secundaria/edad/4esomatematicasB/trigon ometria/impresos/quincena7.pdf
• https://fqm193.ugr.es/media/grupos/FQM193/cms/TFM_(Fco_Javier_Fern andez_Medina).pdf
• https://slideplayer.es/slide/6271883/
• https://es.khanacademy.org/math/geometry/hs-geo-trig/hs-geo-modeling- with-right-triangles/e/applying-right-triangles
• https://www.vitutor.com/al/trigo/tr_e.html
• https://cuentosymates.blogspot.com/2014/02/catetos-e-hipotenusa.html
• https://naukas.com/2011/11/28/ingenio-egipcio-o-como-adelantarse-a- pitagoras-tying-12-knots/rope-12-knots/
• https://www.studocu.com/es-ar/document/universidad-de-buenos-aires/matematicas/alsina-claudi-claudi-apuntes-1-2/9878735

Printed by Books on Demand GmbH, Norderstedt / Germany